What Every Inventor Ought to Know About Designing, Engineering, and Mass Manufacturing Their Idea

by Rob Gramer

Warning-Disclaimer: I wrote this report to provide information in regard to the subject matter covered. It is offered with the understanding that the publisher and the author are not liable for the misconception or misuse of the information provided.

Every effort has been made to make this report as complete and accurate as possible. The purpose of this report is to educate. The author and the publisher shall have neither liability nor responsibility to any person or entity with respect to any loss, damage, or injury caused or alleged to be caused directly or indirectly by the information contained in this report. No offer of investment is being made. The information presented herein is in no way intended as a substitute for legal counseling.

Inside:

The pros and cons of manufacturing in the U.S. versus China.

The difference between designing a prototype and designing an end product

5 ways to use a prototype (and why it's better if the first one is NOT perfect)

How much you can expect to spend on mass manufacturing and the #1 rule to reducing costs

How long most big manufacturers spend on prototyping

What Mark Smith learned about manufacturing in Asia after multiple trips there (when you should consider manufacturing in China)

The step-by-step process from going to idea to end product (and why the first step IS NOT making the thing or protecting it with a patent)

How to determine if someone is qualified to build your idea (and what you must ask for FIRST before you hire them to build it)

How to keep your idea secret while hiring someone to build it for you

An easy rule of thumb for marking up your product to ensure you make a profit

Introduction:

In this interview, you're going to hear from an expert on prototyping and mass manufacturing.

His name is Mark Smith.

And he is going to reveal what he has learned in his past ten years of taking ideas from paper...to prototype...to mass manufacturing...to selling millions of units per month.

It doesn't matter if you've never built a thing in your life...the secrets shared here will work for all different kinds of inventions.

The purpose of this interview is to help show you how to create your idea - either one prototype or one million manufactured end products - as quick and as cheap as humanly possible.

Rob Gramer: So anyways Mark lets kind of start this of by, why don't give me your background, how you got into engineering and I know we went to college together.

Back then you were working for, I think the firm that you worked for couple of years afterwards doing design, engineering, and prototypes. Why don't you give me the short story?

Mark Smith: Okay, got into engineering because I love tinkering, I love math, I like taking stuff apart, I want to know how to take things apart and put them back together.

I decided mechanical engineering because of my love for materials and the tangible parts of engineering.

My first job at college was working with in aerospace, actually for satellite design for rugged space environments. (**Rob Gramer's notes:** designing objects for space is waaaay tougher than anything here on earth. You've got crazy weight requirements - because these things need to fit in a certain payload for takeoff - and all the materials need to work in zero gravity and super harsh conditions.)

So if you can design up for space, you can design for anything.

And so that's how my career kicked off from electronic packaging/product design standpoint.

From there went to more defense work for about 8 years and then went to a startup company where we did commercial and defense products and that's where most of my prototyping and design experience came from.

That company was called G5 Engineering and we did a lot of consumer electronics, a lot of aerospace and very complicated products with lots of different disciplines involved.

So a lot of that experience pretty much allowed me to view any product or any device from how it will be designed, how will be prototyped, how will be tested, marketed, etc.

From there I got my MBA, Master of Business, which also allows me to see the product from a more holistic standpoint, you know more than just design but now we're talking about design efficiency in terms of controlling cost and offering value to the customer and how much you should charge, cost targets, you know, all the different aspects of the life cycle of a product...to create something from scratch and be able to visualize and take it all the way to the product offering.

Rob Gramer: And all the people listening to this interview, at least from my experience, they are inventors.

They are people who have an idea, kinda want to figure out...but don't have the technical background of engineers like us.

So one of the things I wanted you to hit upon here was building a prototype.

Can you explain the difference in building or designing for an end product?

What a prototype is for?

Can you talk about that a little bit like what's the difference between making a prototype and making an end product?

Mark Smith: Sure. Whenever you have an idea of any kind, and it gets to any level of maturity...maybe you've designed it or you know what it should do or you know how to make it do what you want it to do. First thing you want to do is prototype that design.

(**Rob Gramer's notes**: before you ramp up to thousands of units, you want to make a bunch of prototypes to see how to make it better/faster/cheaper/etc.)

So, you want to answer how it works? Maybe there's some special gears or special linkages inside a product that you need to verify? Maybe you want to see materials and finishes but you don't necessarily care about the materials being as strong as it needs to be and real life? These are factors needed for production, but not for prototyping.

So you go through a prototyping process which is really just a functional model, something that is not necessarily ready to sell. Just a proof of concept.

Rob Gramer: So that's a good definition of a prototype. You are not really making something to sell. You are making something more to kind of learn to what's going to happen...you are making a prototype so you can play around with it.

You may say wow, I need to get another screw here or I need to get a smaller piece of plastic or another link here. The prototype is a learning tool and not so much something that you are going to sell to somebody.

Mark Smith: Exactly. Sometimes you are trying to market it to people - like you might take it to a customer show and tell - but in the end prototyping is really just getting something done quickly and efficiently. It will help you solve an unknown or some shortage with that product like. Something with looks, something with the desired ergonomics perhaps.

So that's where a lot of these prototyping processes excel: being able to make something in a week or two weeks or sometimes less depending on what the goal is.

Rob Gramer: Yeah. You hit upon time like a week or two weeks.

Is that something that a lot of your clients, you know they want to see a prototype fairly quickly or they don't want to see quickly. Why did you mention the one or two weeks there?

Mark Smith: It's pretty much exactly what you just said. It depends on the priority of a customer but a lot time these prototype process is going into a greater life cycle or greater design process where the product design process might take six months to a year.

And if you have to wait three months for tooling to get a prototype...that may extend the entire process by months.

So you want to get a process there that can create something for you that's fits in overall time line of the overall process.

So, if the delivery milestone is six months out, you may need to spend three months or so just to make a tool. So you have to factor that in.

That makes a timeline very important.

And there is also the cost factor. It really is the most important for most customers. So they want to get something efficiently made. If I spend a lot of time on the prototype...I know about all of these factors, and I can give them an answer without breaking their bank early on.

So they can get something efficiently made.

Rob Gramer: A lot of the times when I work with the inventors, you know, you get somebody who was tinkering with their prototype, tinkering with their design for years and years on end. Or even just drawing it without really getting to the point that they are actually making something.

You got to make something. If you don't make something, you are not going to have a finished part. And if you don't have a finished part, you'll never have anything to sell.

So I think that's a really important to people listening to this interview is that speed is one of those factors that you want to really become friends with because the faster you are...the faster you learn from something...and the faster you learn from something, the faster you'll get to a sellable product.

Let's switch gears here...

So last time we talked, you said you had gone to, you were in Taiwan, you were in China right?

Mark Smith: Right, Taiwan yeah.

Rob Gramer: Ok. So I know a lot of people I talk to, they hear about China, they hear about manufacturing in China and all these sort of things.

What's something that, you know, let's say somebody is in the production phase and they want to make 100,000 units and they think they need to go over to China.

Why would they want to do that?

Why would they want to stay away from that?

What are some the things that you've learned by your trips over there dealing with these guys on a daily basis.

Mark Smith: The main efficiency or advantage gained by going to China is cost, hands down.

So, you are going to get something made in China, say it's a plastic part may cost, $30,000 for tooling in America, then have some set labor rate ,where you can go to China and get that same part made for probably $5,000 tooling and then probably a tenth of the labor.

So that's the biggest reason you would go to China.

You know, past that, there's not really an efficiency gain there.

If you want to make something that's manufactured on China, it's usually because you are trying to be positioned for lowest cost in the market.

(**Rob Gramer's notes**: I missed this in the interview but it's a key point. And that is...what is your objective in the marketplace? If you want to be cheapest then YES you must manufacture in China. But you may not want to be the cheapest. Take Rolls Royce. They are known as one of the most expensive cars on the planet. They don't WANT to be cheapest. So going to China, and manufacturing there, provides them absolutely no competitive advantage. So before you run off to China ask yourself this question...am I trying to compete on price?)

If you are trying to make a widget or a tool that you are in a big competitive space for with - say like competing with Toys'r'us or one of the big toy manufactures, that's really almost compelled to go over there just to get the cost advantage.

Rob Gramer: So this is something that's really comes into play later on…. It's not like a prototype thing or even a first round thing if you are selling 100 units or even 500 units or maybe much if you don't have to make the trip over there.

It's really something to say okay and now that we are going up to economy to scale or we need to make this special tool that makes us special cutter or something along the line that's only when I want to go over there.

Mark Smith: Right, when it's time to ramp it up for production and that's your biggest cost...that's where you want to go.

Rob Gramer: I went out to Arizona last year and I went to go tour the factory where Ruger manufactures handguns.

So they are popping out one handgun every 37 seconds. And they sell them for about $400 dollars or so. Guns are pretty competitive in the marketplace.

So Ruger is making this fairly complex mechanical thing, that they are making at competitive prices here in the U.S., and they are doing mass manufacturing.

So I would even say to most people that you don't worry about China at the beginning.

It isn't something that even factors in.

Mark Smith: That's true. And the main thing is about how it fits in the marketplace. So if it's a complex part, China has experts like we do but the barriers that you create by going to China are a factor. So if there's something complex that you have a lot invested in from a knowledge base standpoint where it's something that you only know how to do, you may have an issue of communicating it.

And then when it comes to making the tooling part of it, you need to be there to kind of manage the tool, how it's made or any set up that you need for the project.

But if it is something that can be done and has been done efficiently in China then you could just hand it off.

That is, if you are comfortable handing off to somebody. And then they could manage it and make sure that you get a quality product...then that might be a better candidate to take to China.

Rob Gramer: Ok. Let's talk about project management for a second. You know, I am an engineer, you are an engineer, bunch of our friends are engineers so we are kind of inherently get this.

But for somebody who is a lay person, never gone through this process...let's say I have an idea and I woke up in the morning and I say this is a great idea.

If I want to, if there is a step by step to this, you know, what are those steps?

What's the first thing I want to do, what's the second thing I want to do and what's the third thing I want to do and so on and so forth.

Mark Smith: Ok, I can just give you a high level here because I can probably talk for three hours on all the steps needed to take a project to completion.

So let's settle in with a cup of coffee and...

At the beginning when you wake up with the idea, you want to draw the design, you want to brainstorm, you want to talk with your friends, you want to see where this idea is going to go.

Basically data collection.

And so you want to get all the pieces together...all the great ideas, all the stupid ideas, everything. Get it together at once and then start figuring out exactly you want to do with that product. And where it's going to sit in the market.

It's all about making sure it's actually feasible at the beginning.

If it's not, you know, why even do it? If it can't sell, you don't want to bother with it.

If it's just a hobby, that's a whole different issue. I am talking about how to get something to market that's going to be viable and profitable.

Rob Gramer: So even before the engineering phase you are talking about the marketing phase, if there's a market for it and what that market is willing to pay for your idea.

Mark Smith: That's exactly right. You have to look at how it fits...everywhere and for every person who could possibly be interested in buying your product. You have to do this if you want to be successful.

I am not saying that you can't be successful because you love your idea so much that you know that products going to take off. If it's revolutionary and it doesn't matter how much it cost, you know that's a different story.

But even then you still need to plan a little bit to avoid the surprises.

But much of the time, somebody may have already thought about it anyway so there's going to be patent searches.

Something like that has to happen up front before it makes sense to ever go forward with an idea.

But in essence you are absolutely right. You must make sure you even have a chance of doing it before you start because you may just end up wasting a ton of time and money.

Rob Gramer: To sum up the idea phase...you want to get all your information together. You want to look at it from a marketing standpoint. Ask yourself, is this something that the market is going to pay for? From the patent side, find an attorney to see if somebody has any sort of rights with anything relating to your idea. That would be sort of like step one from high level?

Mark Smith: There's probably another 40 page book of details, but in essence that's what it is. You should get your ducks in a row before you take off.

Next would be the prototyping phase. Now you actually start making what you want. If you have that experience and that's something you could do on your own, great. Draw it up and bang it out.

But a lot of the times of this stage you are going to need some sort of engineering input, some sort of design input.

Because you have to get this product made at some point, you are going to need to have it in the right format. That may be 3D software or whatever it may be for your particular product...somebody is going to have to do the work if you can't.

And that can be a huge physical barrier. Just being able to communicate with whoever is going to make the product, right?

So you may hire a prototyping house. And you'll talk to them about refining the idea...design...prototyping...and they will do all that for you, as long as you pay for it, obviously.

But you have to evaluate what they do, just to make sure that it fits within everything that you bid on the first place and everything is still kind of synced up for that product.

Rob Gramer: Right and getting back to what we talked earlier, this prototype can be anything from, you know, you are going to test it out just to make sure it works...it may be something that you will use on potential customers as a testing tool... it or may be a prototype that you take to investors and say hey listen, this is my idea...it's still a work in progress, but here's why you should give me money to go forward with it.

So a prototype can have all these different sort of uses depending on what is the next step is in your plan.

Mark Smith: Exactly right.

And time is going to be big factor in this phase. So if somebody takes this to a design house or whoever you design this with, whether you do or somebody else does it, you want to know exactly how long it is going to take them.

Obviously it would vary by project. But most of the time you spend too much time designing something then you don't want to wait too much for it to be made.

A lot of things can be made in a few days...or weeks...but because of a manufacturers schedule, you may have to wait months before they can even start.

Rob Gramer: Yeah, Personally I have made plenty of prototypes in a few days, or even hours.

But that's because I have the background. I know I can just drive down to home depot, and then spend four hours making something that it would prove the concept.

I can hold it and I can play with it and see whether or not it's something I want to move forward from.

And that's, everybody listening, that's what I want to really get across here...

When you find somebody, who knows this sort of stuff, you are talking about the difference between a prototype that takes three months and a prototype that takes three days or three hours.

So if you find the right person, you find somebody who can do in a timely manner. Explain everything upfront. This will save you a TON of time and frustration right out of the starting gate.

Right, so step one is planning. Doing a little bit of design...patent searches...and market research. Second step is prototyping.

So what comes next?

Mark Smith: It varies just a little bit here depending on funding. If you don't already money, you have to go into a capital acquisition phase. So that's going to be a variable.

Passed that, once you do the business case for your prototype, make sure that it still fits on the market.

Then - when you make it - make sure it fits within the cost target that you want to make it.

Beyond that, for engineering, there are structural tests, and then a static customer review.

I mean whatever you do at this stage make sure that your product still fits on in the market like got in the beginning for just your plan accordingly.

The next stage is to actually start with the detailed design and that's where you start engaging with vendors and suppliers. If you decide to go to China, that's the time to start talking with suppliers and now you are going to start the detail design portion where you pick each one of those individual components and actually start making them for mass production.

So if there is any metal components you need to decide if you want die cast them or machine them, you have to decide about these processes and find whatever the cheapest and most efficient process is for you.

Same thing goes with plastic and all the standard processes for making those parts.

So this is where you really start nailing down the details of the design and figuring out how and where you want to manufacture.

Rob Gramer: And I think we had a conversation of this before..about appropriate cost and how to turn a profit. And we should probably talk about this now.

And just so everybody listening to knows generally I like to look at something that has a markup of at least 5 times, minimum 8 times of that.

If you are spending a dollar on something, then you need to sell it for $5 for a 5 times markup. Or better yet $8 for an 8 times markup. Any less than that and it is going to be really hard for you to make any money.

I don't know if you agree with that Mark Smith or if you have any experience on that, it is the kind of number that I look at.

Mark Smith: Yeah, it's a good adage because it takes into account money without really getting into details.

It takes into account the distribution side of it and if you are going to retail things versus wholesale. A mark up of 5 times is a good short hand way of saying it. A good short hand way of ensuring that you are going to make a profit and then everybody out there on the chain also going to make a profit.

So it's going to vary a little bit with your business model as far as how do you want to sell it, if you want to direct sell or go retail or whatever you choose.

But that's something you really need to work out as a business case side of thing to make sure that you have the right target.

Sometimes, five times is going to price you out of the marketet. And that idea may not make sense to pursue. You just have to figure that out what you are selling, who you are selling it too.

Rob Gramer: Yeah, and we are circling back to market research.

When I deal with inventors for the first time I tell them, you have to know how much it is going to cost to make it...how much it is going to cost to market it...what your sales prices is, etc.

But all these things can paralyze somebody.

If you say listen, whenever it costs to make the thing...you need to sell it for 5 to 8 times markup to cover your costs.

Mark Smith: Especially in the beginning stages you don't want to get bogged down in over analysis.

Rob Gramer: Right. On a related note, we were talking about economies of scale earlier.

You said sometimes in the engineering side, it pays to figure out how you can use off the shelf parts. To save money, because if you are talking about something costing 10 cents versus something costing 12 cents, well now you make a 100,000, that's a huge price difference, correct?

Mark Smith: Absolutely, everything scales up.

So if you can save money on individual components obviously the sum of that would add up quick.

So you want somebody that does this for a living to get you that simple perspective. You want to know how to replace complicated, expensive parts with simple standard parts that would give you the same solution.

This is where engineering really comes into play. How can you build the best product the cheapest.

Rob Gramer: Awhile back I was building a prototype for an inventor I work with. I was at a buddy's shop who builds offroad vehicles. He basically welds and builds stuff all day long.

So we got the design and prototype together and I brought it to him.

Actually I was about to build 4 or 5 or 6 of these things so she can start the prototype and testing phase.

Anyways, he looked at it immediately and he said well, you got this, you got this and you have this, this, and this.

You know, you could just as easily used a bolt with a couple of nuts to hold everything together.

So I ended up starting over from scratch. And the $10 prototype got down to $2.

Have you ever had any experience with that or have you ever heard or seen anything like that in your working with bigger companies and bigger prototypes and stuff like that.

Mark Smith: Almost on a daily basis. Here where I work now. We deal with products that ship 14 million units a year. So when we can save a penny, that's a hundred and forty thousand dollars so, you know, that pretty much pay's a couple of salaries of the people that work in that project just...by just saving a penny.

So it's, when everything scales up and that's really how you have to look at it. You have to take a holistic kind of approach again.

Like when you sell a million units, the little saving add up here and there.

And at that point in becomes imperative to control the cost on each individual component to try to maximize profits.

That's going to leave you open to do other things later on with that money that you saved.

Rob Gramer: So Mark, a lot of good information on this call. I appreciate it. If anybody listening wants to get hold of you, what's the best way for them to reach you?

Mark Smith: Yeah you can get a hold of me by email and phone. My email is MarkSmith7381@gmail.com and my cell phone number, is 850-443-9555. I will be happy to help any way I can.

Rob Gramer: Perfect. All right. And so anybody listening to this interview wants to get a hold of Mark Smith, contact him via email or phone.

And just to verify this again, they can contact you for prototypes, mass manufacturing and any of that sort of stuff right?

Mark Smith: Right, anything from initial concept to taking an idea to production, including a business case scenario. How much you should charge...how many you need to make to break even, things like that, I can help you with that as well.

Rob Gramer: A lot like a business plan.

Mark Smith: Yeah the whole business side of it, I can definitely help.

Rob Gramer: Ok, Perfect Mark. Give your email and phone number one more time and then we will call it a day.

All right, cool. Email is Mark Smith7381@gmail.com and the phone number with area code is 850-443-9555

Awesome. Thanks Mark. I appreciate it.

Have an Idea But Don't Know Where to Start?

I can help you out with that.

From skateboard ramps to jet engines, I've been designing and building things since I was five years old.

Just shoot me a line at rob@inventionprep.com. Give me your name and phone number, and I'll follow up with a phone call to chat about helping you make that thing in your head something you can hold in your hands.

And if you're not at that stage just yet, I can also show you:

- How to save thousands of dollars on legal fees while patenting your invention
- How to find designers and engineers to create your ideas on paper and in real life (from sketches to prototypes, to mass manufacturing) quicker than you ever dreamed possible

- And, how to get all the money you need WITHOUT investors...usually within 30 days (AND you get to keep ALL your equity)

Most people think it takes a major investment of time and money to get their ideas off the ground.

Now you can start profiting from your ideas and inventions in as little as 30 days.

If you'd like me to help, just send an email to Rob@inventionprep.com and we'll take it from there.

Awesome Free Bonuses!

Visit www.inventionprep.com for free instant access to more free cool stuff like...

- Personal feedback from licensed patent professionals, engineers, and experienced marketers on how to protect, create, and sell your idea.
- How to save thousands on legal fees to protect your idea
- Quickly and cheaply create prototypes and final products (in days instead of weeks or months)
- Profit from your idea as quick as humanely possible...usually inside 30 days

Just go to www.inventionprep.com for instant access.

www.ingramcontent.com/pod-product-compliance
Lightning Source LLC
Chambersburg PA
CBHW071543170526
45166CB00004B/1534